PUBLISHED IN 2018 BY
KODOMO PUBLISHING

COPYRIGHT 'ILLUSTRATIONS' 2018 KODOMO PUBLISHING
ALL RIGHT RESERVED. 'NO PART OF THIS PUBLICATION MAY BE REPORDICED OR TRANSMITTED IN ANY
FORM OR BY ANY MEANS. ELECTRONIC, OR MECHANICAL,INCLUDING PHOTOCOPY, RECORDING OR ANY
INFORMATION STORAGE SYSTEM AND RETRIEVAL SYSTEM WITHOUT PERMISSION IN WRITING BY
PUBLISHER 'KODOMO PUBLISHING'

PRINTED IN THE UNITED STATES OF AMERICA

COLOR TEST PAGE

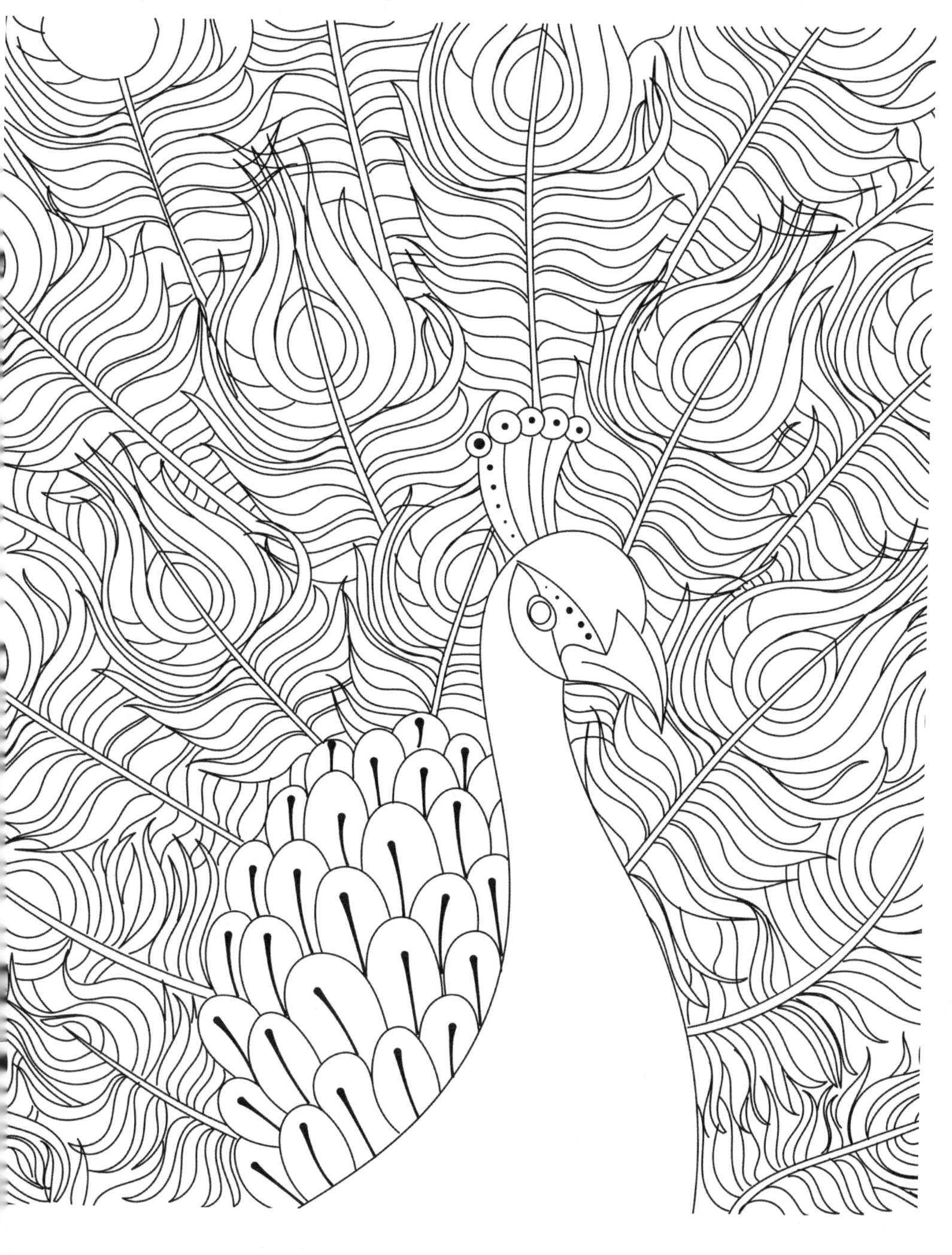

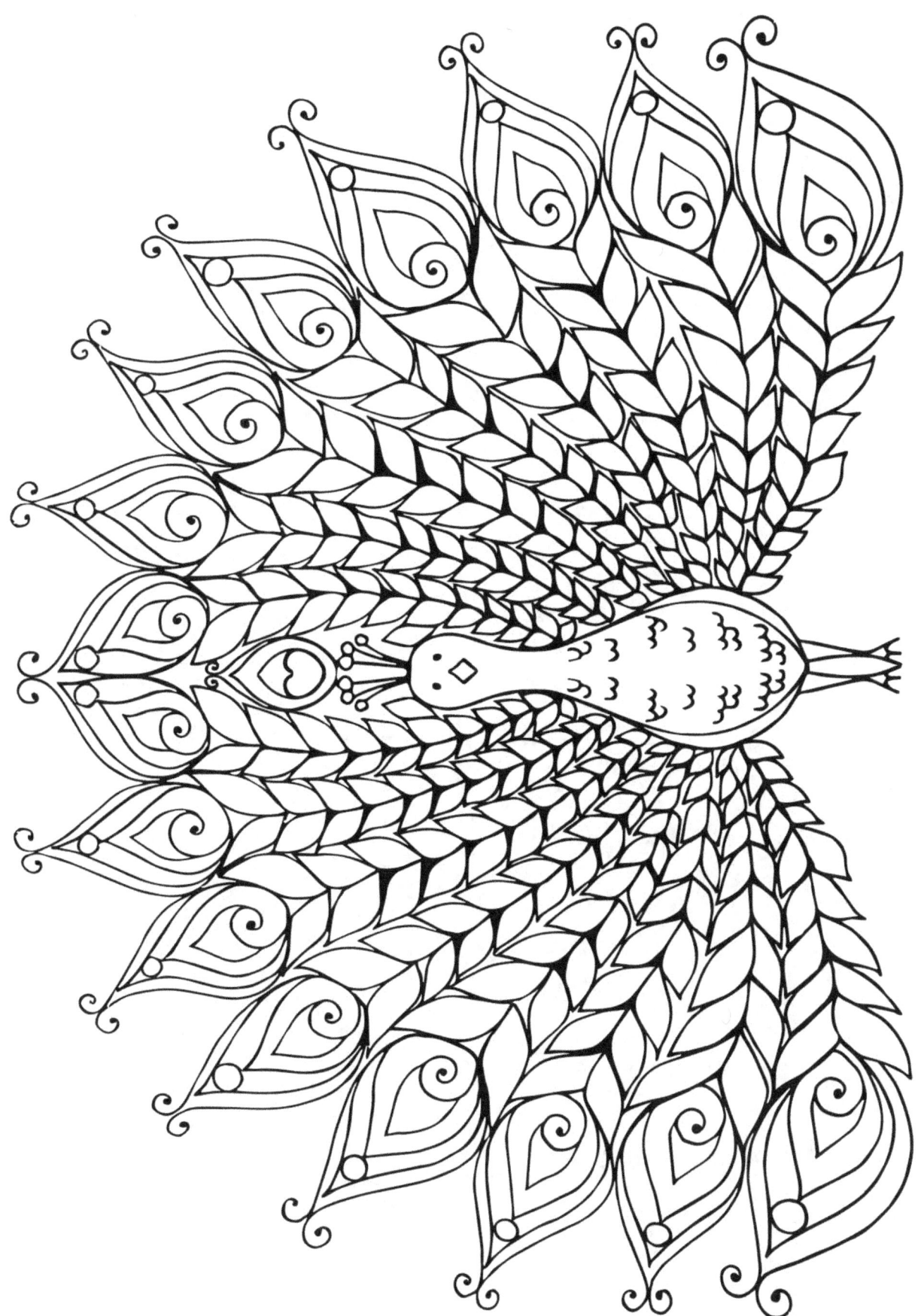

www.ingramcontent.com/pod-product-compliance
Lightning Source LLC
Chambersburg PA
CBHW081018240526
45471CB00017B/3281